AMAZON ECHO IN 1 HOUR

Amazon Echo in 1 Hour:
The Complete Guide for Beginners - Change Your Life, Create Your Smart Home and Do Anything with Alexa!

Joel Goodwin

Joel Goodwin
2017

Copyright © 2017 by Joel Goodwin

All rights Reserved. No part of this publication or the information in it may be quoted from or reproduced in any form by means such as printing, scanning, photocopying or otherwise without prior written permission of the copyright holder.

Disclaimer and Terms of Use: Effort has been made to ensure that the information in this book is accurate and complete, however, the author and the publisher do not warrant the accuracy of the information, text and graphics contained within the book due to the rapidly changing nature of science, research, known and unknown facts and internet. The Author and the publisher do not hold any responsibility for errors, omissions or contrary interpretation of the subject matter herein. This book is presented solely for motivational and informational purposes only.

First Printing: 2017

ISBN 978-1544181936

Contents

Introduction 1
Chapter 1: Your New Best Friend 3
    Know your Echo 4
    The sweet internals for geeks out there! 5
    Learning how to start up your device 7
    Chapter 2: The core features of Echo 11
    Firing up your jams! 11
    No music service? Go old school with radios! 15
    Tune into your favorite audiobooks 16
    Stay updated with the latest traffic reports 18
    Save yourself from a storm 18
    Know which team is winning! 19
    Alexa as your personal wakeup caller 21
    Use your device as a nifty ticker timer 24
    Set up a ticker timer 25
    Create a bond between your Google Calendar and Echo
    27

Chapter 3: More intricate secret of your Echo 29
    Using the device as your Ultimate Bluetooth Speaker 29
    Your custom shopping list! 30
    Have a big Family? No Problem! 33
    Change the way you call your device! 34
    Easily delete all of your recordings! 35
    Just enjoy a few silly games 36
    Know the exact location of your Amazon Packages 37
    Your very own kitchen assistant in the form of Echo .. 37
    Using Echo as your personal calculator 37
    Echo's got skills! 38

Chapter 4: Time to automate your home 43
    The IFTTT is the magical secret! 43
    The list of compatible devices 44
    An alternative way to add your devices 51
    The Final Step 52

Chapter 5: A brief troubleshooting guide and FAQs 54
    Lost Wi-Fi connection 54

Unable to find connected home devices 54
Soft-resetting your Echo 55
Echo not being able to hear properly 56
FAQ Section 56

Conclusion 60

Joel Goodwin

Introduction

Nowadays, it seems like everyone out there is simply trying to turn their house into the next futuristic masterpiece, and for that reason, people are running straight out of their doors and into shopping markets to purchase expensive gadgets to make their home "automated."

Keeping that in mind, in 2014, Amazon released the Echo, which was undoubtedly a giant leap toward the future for mankind and brought them one step closer to affordable home automation. Unsurprisingly, the device turned into an instant hit and won the hearts of millions out there.

Even though the device has been primarily advertised as being a simple Bluetooth Speaker, that is not the only feature the device is designed to have. Thanks to the advanced companion integrated into the device's guts, codenamed "Alexa," the device is able to follow up on your every command with just a whisper of your voice. This is also not to mention the thousands of apps available for Alexa, dubbed as "Skills," which only helps make the device the next "cool" thing around the block.

When Alexa was first released, it gained an overwhelming positive response whereupon the launch of the device has been compared to the innovative breakthrough that occurred when Apple released their first iPhone. With Alexa, nothing was out of your reach anymore. Want a pizza, a cabbie or your latest news and weather report? Everything was achievable through Echo with the help of Alexa and subtle touches of your voice! And that's not even considering the capability of Alexa to actually allow you to control all the smart gadgets in your house!

This is as close to the future as you are going to get.

Even though Amazon did provide a nifty instruction manual alongside the device, it still holds itself way back, which is why newcomers often face some roadblocks.

This is the only book you are ever going to need to fully understand what your new device is capable of and help you unlock the true potential of the Amazon Echo.

Amazon Echo in 1 Hour

Stop wasting any more of your precious time and figure out how Echo might be able to completely alter the whole outlook of your life and plunge you into the future.

Chapter 1: Your New Best Friend

With the holiday seasons incoming, it won't be long before people head out to get the next technologically advanced gadget to show off to their friends and families.

And on the topic of technologies, we aren't far away from plummeting into a future where the boundaries between the interaction of man and machine are blurred. Even today, people are on the verge of fully automating their rooms and houses, which means your home can tell the time, wake you up in the morning and even make your morning toast automated! Technologies are eventually replacing the need of another human being to perform the simple and mundane yet essential tasks of life. Just to let you know, they can even remind you of your mom's birthday!

And all of this is possible thanks to the recent advancements in A.I, or Artificial Intelligence, which is pretty big now. It won't be long before machines will rise up to annihilate us!

But we can always enjoy our time with them until that time comes, right?

What better way for you to jump on the automation bandwagon other than picking up the state of the art Amazon Echo! Be warned, though. Just because at its core it is a Bluetooth speaker, doesn't mean it has to be like that!

In fact, the device packs much more horsepower than you might think.

If you have already purchased your device, then you already know this and are probably excited to jump into the world of the Amazon Echo and explore the myriad of features and functionalities it is hiding, right?

I would like you to slow down a little bit before getting inside the secrets of this device and first let you know how you should set up your device and understand the technology behind your futuristic new gadget!

Slow and steady wins the race, remember?

Know your Echo

So, let us first start our journey by fully explaining what you just bought.

Well, at its core the Amazon Echo has been fully advertised as being a beautiful "Smart" Bluetooth speaker by Amazon. Two things to consider are the "Smart" and "Bluetooth Speaker" parts.

The device sports a nice cylindrical shape with a height of approximately 9.25 inches. To anyone who is not now aware of the secret capabilities of this device, it will just seem like an ordinary yet stylish Bluetooth speaker, but it won't take more than a few minutes for a person to realize there's much more than meets the eye here.

Going back a little bit, the "Smart" nametag of this device comes from the fact that this device has a built-in, internally programmed artificial assistant similar to Cortana from the Halo video game series (we really are in the future). Only here, she is called Alexa. In fact, there are lots of people out there who actually call their device by her name, "Alexa," and you are going to be doing that a lot!

With the help of Alexa, you will be able to fully explore and manipulate all the features of this device by only using the power of your voice! The common abilities Alexa sports include playing your songs, creating a complete shopping list, sending you the latest news and traffic updates and even controlling the lights of your room! We will obviously be discussing each and every aspect of this device throughout the book.

Patience, dear– all in due time!

Joel Goodwin

The sweet internals for geeks out there!

This particular section of the book is primarily targeted toward the nerds among you keen to know the exact mechanism hiding behind the plastic exterior of the device.

Size	9.3" x 3.3" x 3.3" (235 mm x 84 mm x 84 mm)
Weight	37.5 oz. (1064 grams) *These might not be a 1:1 accurate representation*
Wi-Fi Connectivity	The device sports dual banded, dual antennae (MIMO) Wi-Fi, which allows the device to stream much faster while being able to connect Wi-Fi networks under 802.11 a/b/g/n. Be warned, though, that the device is not compatible with peer-to-peer networks.
Bluetooth	The Bluetooth streaming capabilities of this have

Amazon Echo in 1 Hour

Connectivity	been modernized with the Advanced Audio Distribution profile (A2DP for short). Through this technology, you will be able to stream audio to your Echo. Make note that hands-free control with your voice is not yet supported when it comes to Mac OS X-installed devices.
Audio	The device is packed with a sub-woofer of 2.5 inches and a tweeter of 2 inches.
System Requirements	The Echo is a versatile device compatible with most systems out there, and therefore the Alexa app is compatible with all Android, iOS and Fire OS devices, not to mention the features are also changeable via web browser.
Warranty and Service	The device has a 1-year warranty with extra options for another 1-, 2- or 3-year extended service applicable to only US citizens.
Included in the Box	In the box, you will find the device itself alongside the power cable and the usual starter guide.

As you can see, the Echo actually has some pretty nifty specs under its hood.

Learning how to start up your device

Many of you out there might already be an advanced user when it comes to handling the Amazon Echo and are just browsing through the book to explore and get to know something new. But for those of you who are absolute beginners here, setting up the device without any knowledge of how it works might be a little tricky. To make the process much easier and accessible, I am going to guide you through a step-by-step procedure to help you set up the device without any complications.

Even though I already mentioned that setting up the device might be a little bit tricky, Amazon still did a fantastic job in boasting the device to be as user-friendly as possible. The initial process of setting up this device is no different from any other Bluetooth technology you may have used in the past.

Before embarking on the following steps, make sure you have placed your Echo in such a place where it is not facing any obstruction within 8-9 inches of its surrounding radius.

Step 1

The first step is pretty simple: right after you fire up your device and connect your Echo to a power outlet, you will want to download the "Alexa" app from the Google Play or Apple store, depending on your needs.

Step 2

Once you are done downloading the app, next you will want to turn on your device. A nice blue ring on top of the device will slowly turn into a rotating orange color, indicating the device is ready to be configured.

Once the orange color has appeared, take out your smartphone and access your Wi-Fi settings. Echo will directly appear in your wireless device list and will require you to ad-hoc directly to the device to continue the configuration process. Once you are in the list, look for "Amazon 53N," which will indicate the device, as shown in the diagram.

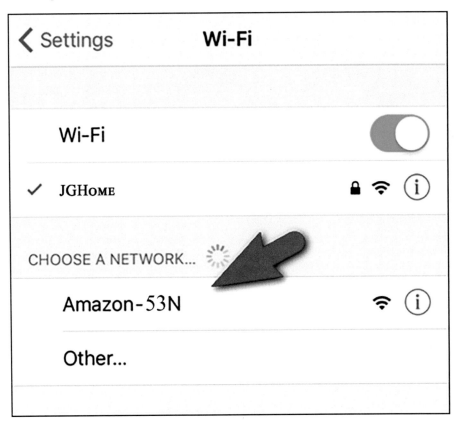

Step 3

Once done, fire up your downloaded Alexa app, which will immediately transfer you to the configuration window.

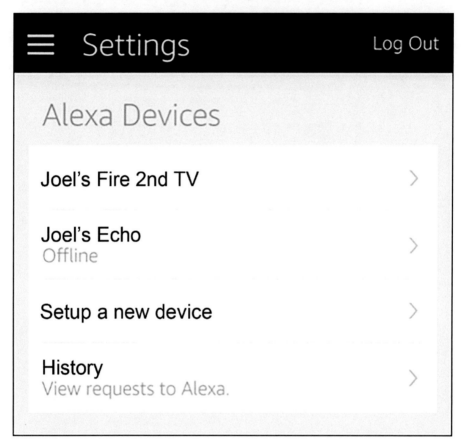

Here, you will be greeted by a screen where you will see the name under which the device was purchased. However, if the device was brought to you as a present, then you will be greeted with another option to "Set up your new device" where you will need to put in all of the desired information and personalize the device accordingly, including the credentials of your Amazon login account. Here, you will also get the chance to read the rules and regulations for using Echo and agree to that.

Amazon Echo in 1 Hour

Step 4

Once that part is complete, next you are going to choose a Wi-Fi network for Echo having an internet connection, which will be used by Echo to perform the required tasks.

Just choose the network from the list, and you're done!

Chapter 2: The core features of Echo

Now that I have taught you how to set up your device fully, we are finally ready to jump into what you've been waiting for! The chapters from now on will show you how to unlock the true potential of this device.

Before going into the more advanced concepts of the device, however, I am going to discuss the basic capabilities of the Amazon Echo. You most certainly should know the core feature of the Echo, or else your friends just might call you crazy! And not in a good way, either.

Just as Aladdin opened up the door to the caves of wonder by saying "Open sesame!" the magic phrase to wake up your device is simply "Alexa."

Whenever she hears you utter her name, she will instantly wake up and prepare herself to listen to whatever command you may have for her.

Firing up your jams!

First and foremost, the Amazon Echo device has been advertised as being a speaker, right? Therefore, the first thing you will want to do is listen to some nice jams using the speaker! This is the moment where you get a glimpse of the immense audio capabilities of your device.

One might imagine that with all of its technical capabilities, playing songs might be difficult, but that's just not the case.

While the media streaming database compatible with Echo is ever so evolving, at the time of this writing, Echo was compatible with the following services, some of which are paid while the others are free:

- Amazon Music
- Prime Music
- Spotify Premium
- Pandora
- TuneIn
- iHeartRadio
- Audible

Now, before you can utilize any of those features, you will need to pair up your device with one of the services, which is pretty easy. Just go to your Alexa App and tap the three bar menu on the upper left. Say, for example, that we want our device to pair up with Pandora, you can click open the menu and go to "Pandora," and then follow the instructions to set up your account.

To do this...	Say this...
Adjust the volume	"Volume up / down." "Set volume to level [number]."
Hear details about the track currently playing	"What is this?" "Who is this?" "What song is this?" "Who is this artist?"
Halt the currently playing track	"Stop." "Pause."

To do this...	Say this...
To play your favorite song	"Play." "Resume."
Manipulate your sleeper timer	"Set a sleep timer for [x] minutes / hours." "Stop playing music in [x] minutes / hours." "Cancel sleep timer."
Change to either the previous or the next track	"Next." "Previous."
Loop the music queue**	"Loop."
Shuffle songs or tracks**	"Shuffle." "Stop shuffle."
Repeat songs or tracks**	"Repeat."
Play Prime Music	"Play [song / album / artist] from Prime Music." "Play some Prime Music." "Play a Prime Playlist." "Play [station name] from

Amazon Echo in 1 Hour

To do this...	Say this...
	Prime." "Play [playlist name] from Prime Music."
Play Spotify Premium **(Compatible only with the Amazon Tap, Echo Dot and of course, the Echo)**	"Play [song name] from Spotify." "Play [song name] by [artist] from Spotify." "Play songs by [artist name] from Spotify." "Play music composed by [composer] from Spotify." "Play [playlist name] from Spotify." "Play 'Discover Weekly' playlist from Spotify." "Play [genre] from Spotify." "Play Spotify." "Spotify Connect. / Connect to Spotify."

Following the same procedure, you will be able to add more accounts as required. Once done, you will want to familiarize yourself with the following voice commands to control the flow of your music.

Joel Goodwin

No music service? Go old school with radios!

Sometimes you might want to go back in time and listen to the radio, going old school to find out what the next hot DJ is about to play! Even though time has passed, the evolution of radio has helped it to stand the tests of time! And by evolution, I mean constantly entertaining services like TuneIn and iHeart, which are winning hearts all around the world!

Radio manipulation is no different from playing your music. Just follow the voice commands listed below to please your heart's desire.

Play a custom station (iHeartRadio, Pandora, and Prime Stations)	"Play my [artist / genre] station on [Pandora / iHeartRadio / Prime Music]."
Play a radio station (TuneIn and iHeartRadio)	"Play [station frequency]." "Play the station [station call sign]." "Play the station [name of the station]."
Playing a podcast program	"Play the podcast [podcast name]." "Play the program [program name]."

Tune into your favorite audiobooks

The advanced voice synthesizing technology appropriately used in creating the technology is extremely sophisticated, which allows Alexa to be used as a fine audio companion who will read all your favorite books out loud! When I was writing this book, Alexa had a strong support for Kindle and Audible, but thanks to the voice synthesizing tech, Alexa can read aloud from just about any book saved in your library, and even from Wikipedia as well!

The following voice commands will allow you to control Alexa's reading capabilities:

To do this...	Say this...
Listen to an audiobook	"Read [title]." "Play the book, [title]." "Play the audiobook, [title]." "Play [title] from Audible."
This option of pause up the current audiobook	"Pause."
Start-up and continue your most recent book	"Resume my book."
Forward or rewind your book read by about 30 seconds	"Go back." "Go forward."
Go to the next or previous chapter	"Next chapter."

To do this...	Say this...
	"Previous chapter."
Go to a specific chapter	"Go to chapter [#]."
Restart a chapter	"Restart."
Manipulate your timer	"Set a sleep timer for [x] minutes / hours." "Stop reading the book in [x] minutes/hours." "Cancel sleep timer."

Stay updated with the latest traffic reports

We've all been there! With the ever-growing number of automobiles roaming the streets, it would be a miracle if someone told you he/she never had to sit on the road for hours on end due to a traffic jam! No more will you need to venture out into the wild without any knowledge of the streets. Alexa is fully capable of providing you with the latest traffic updates within seconds, which will undoubtedly make you the all-seer when it comes to street news.

To set up the intelligent traffic tracking capabilities of Alexa, just do the following:
- Fire up your Alexa app and go to the navigation panel and select traffic
- Change the address and fill in the fields for "From" and "To" (these two locations will determine the route you are eager to be updated about)
- Save the changes

With all of that done, use the following commands:

To do this...	Say this...
Ask for a traffic update	"How is traffic?" "What's my commute?" "What's traffic like right now?"

Save yourself from a storm

Another amazing feature of Echo allows you to gracefully save yourself from another rainy or snow-drenched day! Echo is capable of doing that thanks to services such as AccuWeather.

Once you have set your location, you are good to go. The commands below are going to help you manipulate the intelligent system needed to get to know the status of the weather.

To achieve this	Say this
Get to know the current weather conditions	"What's the weather?"
Get to know the weather of a specifically targeted day	"What's the weather for this weekend?" "What's the weather for this week?" "What's the weather for [day]?"
Get to know the weather of another city of your choice	"What's the weather in [city, state or city, country]?"
Know about future weather conditions	"Will it [rain / snow] tomorrow?" "Will it be windy tomorrow?"

Know which team is winning!

Having a busy lifestyle sometimes forces us to sacrifice our favorite pastimes. While it might be a little difficult to sit in front of the tele to enjoy your favorite sports team go for the gold, that's no reason to get sad, especially not with Alexa here! Using Alexa, you will be able to get the latest updates from your favorite sports team regardless of your current conditions! Even if you are having a good time sitting in your washroom, just shout out for the news on your team!

Features are always increasing, but while writing, Alexa had the support for the following leagues:

- EPL - English Premier League
- MLB - Major League Baseball
- MLS Major League Soccer
- NBA - National Basketball Association
- NCAA men's basketball - National Collegiate Athletic Association
- NCAA FBS football - National Collegiate Athletic Association: Football Bowl Subdivision
- NFL - National Football League
- NHL - National Hockey League
- WNBA - Women's National Basketball Association

To add your favorite sports team, all you need to do is:
- Open up your Alexa App
- Go to Settings > Sports Update
- Punch in the name of your sports team and choose one from the provided list
- Select the team and add it to "Sports Update"

Once you are done with all the formalities, the only voice command you will need to know is "Give me my sports update!" and you are done! Just don't shout out too loudly in excitement.

Alexa as your personal wakeup caller

No more will you need to rely on old-fashioned cell phones or even your wife or mother to wake you up from your slumber! You can just simply hand over the task to dear Alexa. The Echo will give you the option to set up your alarms and manage them with incredible ease! Even if you are completely exhausted from a day of strenuous work, you can easily give the burden of waking you up right on time to Alexa. Without worrying about anything else, this device will allow you to have a good night's sleep and have the dream of a lifetime!

The voice commands required here are as follows:

To achieve this, either use these voice commands	Or set them up through your Alexa app
For setting up one alarm "Wake me up at [time]." "Set an alarm for [time]." "Set an alarm for [amount of time] from now."	Once you have used your voice to set up your alarm, you can easily go to your app and further configure it.
Go for an alarm that will repeat itself "Set a repeating alarm for [day of week] at [time]." "Set an everyday alarm for [time]."	Assuming that you have an alarm set, go through the following steps: • Go to the navigation panel and select Timers • Find your device • Go to the tab named Alarms • Choose your current alarm • Go to the repeats section and select one from the options in the list • Once done, save the changes

Amazon Echo in 1 Hour

To achieve this, either use these voice commands	Or set them up through your Alexa app
Snooze the alarm Say "Snooze" (while the alarm is ringing). **Keep in mind that the alarm will keep snoozing for about 9 minutes.**	Still not configurable.
Check the status of your alarms "What time is my alarm set for?" "What alarms do I have for [day]?" "What repeating alarms do I have?" **Keep in mind that if you have multiple alarms set, Alexa will read out the alarms and let you choose which one you want it to recite.**	• Go to the Timers section • From the menu, choose the tab named alarms • Choose the alarm you are looking for • Choose the option you need
Prevent an Alarm from buzzing up "Stop the alarm" (when alarm is sounding). "Cancel alarm for [time] on [day]" (turns off the alarm, but does not delete it).	• Go to the navigation panel and select Timers • From there, find your device • Go to the tab named Alarms • Find the Alarm you want to disarm and either remove it or delete it

Joel Goodwin

To achieve this, either use these voice commands	Or set them up through your Alexa app
To delete an Alarm Through the app.	• Go to the navigation panel and select Timers • Find your device • Go to the tab named Alarms • Find the Alarm you want to disarm and either remove it or delete it
Alter the Alarm volume Through the app.	• Go to the navigation panel and select Timers • Find your device • Go to the tab named Sounds • Gently press the volume bar and set the alarm volume *Keep in mind that changing the alarm volume won't alter the voice volume of your device
Alter the Alarm sound Through the app.	**For brand new alarms** • Go to the navigation panel and select Timers • Find your device • Go to the tab named Sounds • Choose your different sound and change it Please note that this won't change the sound of existing alarms, but instead it goes for the newer ones. For existing

To achieve this, either use these voice commands	Or set them up through your Alexa app
	ones, follow the instructions below. For an existing alarm: - Go to the navigation panel and select Timers - From there, find your device - Choose the alarm you want to alter - Choose your alarm sound - Save changes upon returning to the former alarm screen

Use your device as a nifty ticker timer

Now, going into a long slumber is super cool and all, but you know what's even cooler? It's taking short power naps and starting your daily work right off the bat! But that's not the only thing we are going for here; imagine that you want to be reminded of something in, say, 10 minutes? Sure, you can go for the good old-fashioned pen and paper, but why bother with such a nuisance when you can simply let Alexa remember everything for you? Don't be that busy man who constantly loses track of everything! Be the smart man who has a personal digital assistant and has everything covered.

So, to tell Alexa to remind you of something, wake you up after a while, or even set a simple ticker timer, you will want to know the following voice commands.

Set up a ticker timer

Alarms are awesome for when you want to knock out for a really long time, but what if you simply want Alexa to remind you of something? For example, you might have to change and shut down your store after 10 minutes. A busy man can easily lose track of time and forget these things! Well, Alexa has got you covered in this department as well by allowing you to set up timers through the following commands:

To achieve this effect, utter these	Alternatively, use your app.
For setting up a countdown timer "Set a timer for [amount of time]." "Set the timer for [time]." Keep in mind that it is not possible to set your timer more than 24 hours in advance.	Just use the power of your voice to set it.
For pausing up or continuing your timer Done through the app itself.	• The first step is to go the navigation panel named Timers and Alarms • From the drop-down menu, choose the name of your device • Go to the tab for the timers • Choose the edit section next to the timer you want to edit and go to pause
Get the information for your	

Amazon Echo in 1 Hour

To achieve this effect, utter these	Alternatively, use your app.
countdown timer "How much time is left on my timer?" **Keep in mind that if you have multiple timers, you will want to get the information through the app.**	• Go to the navigation panel named Timers and Alarms • From the drop-down menu, choose the name of your device • Go to the tab for the timers • Choose the timer and check out the information for your timer
Prevent a countdown timer from counting "Stop the timer" (when timer is sounding). "Cancel the timer for [amount of time]" (for upcoming timers). **If for some reason you have two timers set to expire within the same time, go to the app and adjust it.**	• Go to the navigation panel named Timers and Alarms • From the drop-down menu, choose the name of your device • Go to the tab for the timers • Click the edit button beside your preferred timer
Alter the countdown timer volume Done through the app	• Choose the settings • Choose the device • Go to sounds • Find the volume bar designated for the Alarm and Timer and drag it to your desired level **Keep in mind that the volume of the countdown does not affect the volume of the device.**

Joel Goodwin

Create a bond between your Google Calendar and Echo

Having a busy and intense lifestyle often means you might need to remember a bucket load of different dates in your life. However, it can be impossible to remember all of them.

What if you have a forgetful memory? It never hurts to get help from someone (or something, in our case) to keep us updated, right? That is exactly why Google Calendars was developed! Now, you can easily sync up your Echo with your Google Calendar to seamlessly blend your life.

To do that, follow these steps:
- Open the Alexa app
- Go to Settings and then Calendar
- Select Google Calendar
- Press the Link Google Calendar Account button and enter your credentials

Once done, use the following voice commands to manage the events of your calendar:

To Achieve This	Utter This
Get the information for your upcoming event	"When is my next event?" "What's on my calendar?"
Get to know the exact time of your upcoming event	"What's on my calendar tomorrow at [time]?" "What's on my calendar on [day]?"
Follow these steps to add a new event to your calendar	"Add an event to my calendar." "Add [event] to my calendar for

Amazon Echo in 1 Hour

To Achieve This	Utter This
	[day] at [time]."

And with that you now know the most basic and preliminary actions for your Amazon Echo device.

Now you are ready to tackle the more advanced features of your device.

Chapter 3: More intricate secret of your Echo

Now that you are perfectly clear on the basic features of your device, we are going to look further into the more unusual and lesser-known or advanced features of Echo, which might be new to you at first glance.

Using the device as your Ultimate Bluetooth Speaker

Despite having all these heavy functionalities, Amazon Echo still is at its heart a simple Bluetooth speaker designed to play any music you want at your convenience. Now that we have already pointed out the essential "Smart" features of your device, let us go a bit old school and teach you to use your Echo using just your phone and play music off it. The hassle of having no internet connection won't be a problem any longer, as will you be able to bring the party anywhere you want! This is achieved through the immensely stable Bluetooth technology that allows it to turn into a fully-fledged 360-degree surround sound speaker! The steps are as follows:

- Make sure the Bluetooth on your device is turned on and set to pairing mode
- Bring the device close to your Echo
- Say "Pair" to prepare your Echo to be paired
- Open up the Bluetooth menu on your phone and choose "Echo," then pair it up
- To terminate the connection, just say "Disconnect"

Your custom shopping list!

Yes, you read that right! Using Alexa, you will actually be able to create your very own shopping list, or a list containing all of your life goals! And we all know how much we love to check items off a checklist, right?

To achieve this	Utter this
Creating your shopping list by adding an item	"Add [item] to my Shopping List." "Put [task] on my To-do List."
Go through your list to double check it	"What's on my Shopping List?" "What's on my To-do List?"

With Echo, you won't need to carry pen and paper or even a cell phone to make your list. Just let Alexa know what you want to add to your list and she will take care of everything. What's even better is the fact that, using Alexa, you will be able to print out your list from a connected computer right after viewing it. Using this feature is pretty easy and accessible through the Alexa app. Just use the aforementioned and below voice commands to manipulate the list through the power of your voice!

To perform this action	Do this
Print out your created list	**Using your computer and the Alexa app** • Simply go to your navigation menu and choose the panel title "Shopping & To-Do List"

To perform this action	Do this
	- Select the list you are looking for - Go to print
Open up an existing list	**Using the Alexa app** - Simply go to your navigation menu and choose the panel titled "Shopping & To do list" - Select the list you are looking for to open it up **Keep in mind that having an internet connection is not essential to view your list.**
Add an item to your created list	**Using the Alexa app** - Go to your preferred list - Enter the name of your task/item - Click on the plus symbol
Alter an item from your existing list	**Alexa app** - Go to your list and head over to the item you want to modify - Once found, click edit
Delete an item from your existing list	**Using the Alexa app** - Press the arrow placed next to your desired item (a downward facing one) - From the drop down, choose delete to remove that

Amazon Echo in 1 Hour

To perform this action	Do this
	item.
Select an item and deem it complete	**Using the Alexa app** This is straight forward, as it only requires you to choose the checkbox and click on it to toss in under complete

Have a big Family? No Problem!

Alexa is great when it comes to taking care of different family members. Whether it be a large family or a small one, the taste and musical preference of each and every person might vary. While you might be a fan of deathcore, your sister might be a fan of soft classical music. Therefore, cramming all of your preferences and music stations in just a single Alexa profile might be somewhat of a nuisance. Keeping that in mind, Alex has been designed to blissfully manage different profiles for each house member through the Alexa App. Juggling between these multiple profiles is easy breezy and without any form of complication.

For adding a new profile

As the title of this section implies, the following steps will teach you how to create a new profile for your household.
- From the navigation panel, just go to settings
- Then go to Account and choose the Household Profile option
- Click there and you will be greeted with a number of instructions explaining how to add a second person to your Household
- Follow them properly, and you are done

Removing a profile from your household

Similarly, the following steps will teach you how to remove an individual from your household. However, keep in mind that once you have removed a person, you won't be able to add it back again within the next 180 days. So, be careful of this. If you accidentally remove the profile of someone you were not supposed to, simply contact customer support. Other than that, the following steps must be followed when you want to remove someone:

- From the navigation panel, go to settings
- Go to Account and choose the Household Profile option
- Choose the person whom you want to delete
- Press the remove from household button and the profile of that person will go away

Switch between profiles on the fly

Now that your accounts are all set up, you will next be able to simply go ahead and switch your accounts on the fly in order to access the account of another user. For this, just do the following:
- Utter the words "Switch Account"
- Alternatively, while you are going through the content library in your app, just click on the drop down menu where you will get the options to toggle between libraries

Change the way you call your device!

Is the name Alexa not sweet enough for you? Or is your device not feeling cared for enough when you address her by Alexa? Or is it that your girlfriend has started to scold you because you are constantly uttering the name of another woman? Whatever the case may be, you can just as easily change the wakeup word of your device and assign one that won't send your whole career and love/marriage life spiraling down into chaos!
- Open the Alexa app and go to settings
- Choose the Alexa Device, which should be there under the name of your Echo device
- Click on the option called "Wake Word" and assign your new word.
- Just a small reminder, whenever you have changed the wake word of Echo, it will take some time for the device to properly re-adjust herself and alter the settings according to the newly selected word. During that period, the device will remain unusable and in hibernation, but it's nothing to be afraid of!

Easily delete all of your recordings!

When the device first came out, one of the controversies surrounding the audio-induced nature of the device was that it was able to record and upload everything to the Amazon server. This rumor caused a great deal of backlash when it first circulated, as people feared that this data might be used later on for ill purposes.

While part of it is true, it is also possible to actually erase all of the stored data quite easily, so you won't have to worry about it being released to the CIA!

Just go through the following steps:
- Go to your Alexa App and open up settings
- Go to history
- Here you will see a complete list of all the requests you have made since the initiation of this device
- Just click on the recording you want to delete and it will be done!
- Alternatively, if you want to completely clean the slate, you will need to go to www.amazon.com/myx and sign in there.
- Select your device and go to the "Manage Voice Recording" options where you will have the option to erase everything in one go

Just enjoy a few silly games

Yup, I am not being sarcastic here at all! Alexa is actually able to play a number of fun family games with you! These are good ways to kill some time with your siblings and some fun.

- **Coin Flip:** Using Alexa, you will actually be able to play a classic game of heads and tails! Just simply utter the words "Alexa, heads or tails?" and the game will begin!
- **Bingo:** For the classic family-favorite game, keep in mind that you are going to need to download one of the Bingo cards from the internet. After that, you are practically ready to play a mind stimulating game of Bingo! To initiate it, just utter the words, "Alexa, play Bingo!"
- **Simon Says:** Remember the classic game of Simon Says? Well, Alexa is actually programmed with the software to play Simon Says with you as well. Just go up to and say, "Simon Says" in front of anything you want her to repeat and she will do that promptly. The fun factor of this game amps up if you have an Echo remote with you.
- **Simple Geography Quiz:** While a Geography test might sound a little boring, if you play the game with your friends, this might actually lead to a great time. Just start by saying, "Alexa, Start Animal Game/Capital Quiz" and Alexa will fire up a nice 20-question game of trivia.

Know the exact location of your Amazon Packages

With the recent announcement of Amazon Go, Amazon is kicking it up a notch when it comes to consumer satisfaction. Amazon did include a nice feature of this in Echo as well. Using Echo, you will be able to track the exact location of your delivery and get constant updates. Be aware that this feature still has some bugs, but it works!

- The first step here is to add the link to your ordered package in the Alexa app.
- Next, when you want to get an update, all you'll want to say is, "Alexa, where's my stuff?" and Alexa will immediately update you regarding the delivery of your package.

Your very own kitchen assistant in the form of Echo

Now, obviously, you will be able to use Alexa and make her read the next "Top 100 Ketogenic Diet Recipe Books", but that's not all! Alexa can also help you out in your kitchen by performing basic calculations. These calculate the exact amount of flour or salt you might need for your next culinary masterpiece! The voice commands include:

- Alexa, how many cups make a quart?
- Alexa, how many teaspoons are in a stick of margarine?
- Alexa, set the timer for 10 minutes
- Alexa, convert 15 milliliters to liters

Using Echo as your personal calculator

Yup, Echo is not only capable of acting as your all around entertainment buddy, but it also has the ability to help your kids to carry out simple calculations. The A.I. of Alexa is smart enough to do various calculations, including subtraction, multiplication, addition and even division. In our example, we used different questions, such as, "Alexa, what is three plus two?" or even, "Alexa, divide 444.1 by 666.2."

Needless to say, all of the results were mind-numbingly correct. Alexa is actually able to not only deal with integers but also floating numbers (numbers with decimal points), making this device even more versatile!

Echo's got skills!

Up until now, I have introduced you to some of the more amazing features of your Echo device, but it's not over yet!

To start, Alexa has her very own app store. The app store for Alexa is known as "Skills," which contains thousands of apps that give your Alexa device even more added functionalities. At the time of writing, the skill count was somewhere around 1500+, but it is increasing all the time.

Given the magnitude of the skills available, it won't be possible to cover all of them here in this section, but that doesn't mean I won't touch any of them! Below is a list of some of the best skills out there right now that you should check out immediately!

But first, you must activate your skills:
- Open up your Alexa app
- Go to Settings
- Using the search bar, enter your desired skill or use the "Categories" section to browse through the available skills
- Once you find a skill you are happy with, click enable

As for the list, here we commence:

- **Fitbit:** This is undoubtedly the best possible activity tracker app present in the app store. The app allows you to blissfully keep track of all the various information, such as steps taken, your exercise goal or even sleep tracking to make sure you lead a happy and normal life!
- **The 7-Minute Workout:** If you are into fitness, you should definitely check out this app, as it allows you to keep track of a specific exercise routine, all of which can be followed within just seven minutes or so.
- **Stock Exchange:** This is one app that administers to the budding businessmen out there. With this you will be provided with a plethora of inputs to be inserted, and you will be able to obtain a summary of all the chosen stocks that help you administer them into a larger portfolio.
- **The Bartender:** This is a pretty unique app. Even though the app store is full of ups and downs, this is one of the better ones, undoubtedly! The brilliance of this app allows you to simply tell Alexa to narrate the recipes of all, if not most, of the famous bar drinks out there.
- **Yo Mamma Jokes:** Now, keep in mind that this app is not suitable for all ages, but if you are a person who falls into the "Adult" category, then this will definitely be of interest to you. The app does just as the name implies: it shoots out a large number of "Yo Mamma" jokes whenever you want to hear them! Perfect for a good laugh amongst friends and co-workers.
- **4A Fart:** While the apps in Echo can seem a bit serious, this one is all for the jokes! The app, as its name implies, goes all out to produce horrendous fart noises at your command.

- **The Magic Door:** This is an extremely cool app that will allow you to dive deep into the realm of magic through a grossly enriched narrative. The main purpose of this app is to help your child open their imaginative potential by going through different storylines, all of which are extremely interesting and even come with immersive sound effects and multiple choices to influence the ending of the tale.
- **Akinator:** This is one of the most intriguing games present in the Skill Store. This skill allows you to play a nice game of 20 questions with Alexa, where Alexa has to guess the character you are thinking about!
- **Amazing Word Master Game:** This app allows you to enhance your vocabulary through a game. This app will simply throw random words at you and allow you to give the meaning of them and train you while you are at it.
- **Pick-up Lines:** Are you so hopeless and desperate in your life that you always have to reach out to your friend for life advice on how to impress the girl you love? Well, you won't have to embarrass yourself ever again, as Echo has got you covered. This app will simply keep throwing different pick-up lines at you until you turn into a complete pro.
- **Capital One:** An app sponsored by the Capital One Bank, this app actually allows you to control and get all the credentials of your Capital One bank account. With this, you will be able to check for information on transactions, as well as pay credit card bills with your voice.
- **Campbell's Kitchen:** An app from one of the well-known and trusted food companies out there! The app produces a wide variety of recipes that ranges from soup, meat or even fish. If you are a budding/aspiring chef, you don't want to miss out on this.

- **TV Shows:** The function of this app is pretty much self-explanatory. If you are a TV hog, then this is the ultimate skill for you! The app will provide you with all of the accurate information regarding your favorite TV series and fire at you all of the necessary information, such as the timing of the next episode.
- **Automatic:** With Google leading the industry when it comes to creating automated vehicles, this app is Echo's first step into this world. The Automatic skill allows you to sync with your vehicle pretty nicely and provide you with different stats, such as distance driven, location and gas.
- **Uber:** The world-famous taxi calling app is also available in your Echo as well! As you may all know, Uber helps to catch a ride, but now, you will be able to do it by just uttering a few words. Hire a ride, check your ride status and even check information on the ETA of your hired ride.
- **1-800-Flowers:** Sometimes you might want to surprise your fiancée, your mother or your wife with a beautiful bunch of roses, but our busy lifestyle sometimes prevents us from buying them in time. Well, using this app, you will be able to make your arrangements easily. Just poke your Alexa, and this skill will help you to choose the flowers, your destination as well as the chosen date of delivery.
- **Domino's Pizza:** Definitely one of the leading Italian restaurants out there! With this app, you can simply order your cheesy and yummy pizzas in just a moment's notice. The device will even alert you whenever your pizza has arrived.

- **The Wayne Investigation:** Similar to the Magical Door app, this skill also focuses on creating a large and engrossing story through simple narrative techniques. While the Magical Door was for children, this one is for a matured audience. Using this skill, you will able to play out this game, but here's the catch: you play as the world-famous detective Batman to solve an interesting case. This one also has multiple endings, and sound makes the experience much more immersive. It should be noted that this is amongst the top rated games in Skills.
- **Jeopardy J6:** Any game show enthusiast would easily recognize the name of this game in a jiffy. This app is based on that very show, and no other app does trivia better than this one! Not only will you have fun while playing this game, but you will also get a boost in your general knowledge as well.

Chapter 4: Time to automate your home

This is where the "Smart" part of Amazon Echo comes into play. After reading this chapter, you will know exactly how much potential your device has and how to utilize it.

Besides all the gorgeous services and skills available in Echo, the device can also be used to completely allow you to manipulate the automatic electronics and gadgets in your house in a jiffy. Just have a look at the Jetsons because that is the future you are headed toward with this tech. And the best part? All of your devices can be controlled by only your voice! How cool is that?

Use Echo as the centerpiece of your automated home. You will be able to control the lighting in your house, the room temperature, security panels, and power functions of specific electrical components as well. Interested in knowing how all of these work? Well, I am going to do that just now!

The IFTTT is the magical secret!

The reason why Echo can break down all the barriers between human and electronic interaction is nothing more than the advanced integrated service known as the IFTTT. While the name may sound just a little bit weird, this is actually the software powering the interaction system of the device.

The service is essentially handcrafted from a number of different formulas, which helps Echo to understand human interaction and get to know how it should communicate with another device or service. For example, the IFTTT allows Echo to connect with a Belkin WeMO switch in order to let you manipulate the power level of your lamps.

The list of compatible devices

Now before actually moving onto teaching you how to add the various devices, I thought it might be helpful to give you a quick list of most of the devices compatible with Echo up until the latest update. This should help you start up your journey and take baby steps toward home automation.

Smart Home Hubs

- Samsung SmarThings Hub
- Wink Hub
- Insteon Hub
- Alarm.com Hub
- Vivint Hub
- Nexia Home Intelligence Bridge
- Universal Devices ISY Hubs
- HomeSeer Home Controller
- Simple Control Simple Hub
- Almond Smart Home WiFi Routers

Lighting

- Philips Hue White Starter Kit
- Philips Hue White and Color Ambiance Starter Kit
- Philips Hue Go
- Philips Friends Of Hue Lighting Bloom
- Philips Friends Of Hue Lightstrip
- LIFX Color 1000 A19 Smart Bulb
- LIFX Color 1000 BR30 Smart Bulb
- LIFX Color 800 A19 Smart Bulb
- LIFX Color 900 BR30 Smart Bulb
- Cree Connected LED
- GE Link Bulb
- OsramLightify Smart Bulb
- TCP Connected Smart Bulbs

Switches, Dimmers and Outlets

- Belkin WeMo Light Switch
- Belkin WeMO Switch
- Belkin Wemo Insight Switch
- iHome Smart Plug
- Samsung SmartThings Outlet
- TP-Link HS100 Smart Plug
- TP-Link HS110 Smart Plug With Energy Monitoring
- D-Link WiFi Smart Plugs
- Insteon Switches, Dimmers and Outlets
- GE Z – Wave Switches, Dimmers and Outlets
- Leviton Switches, Dimmers, and Outlets
- Lutron Caseta Wireless Switches, Dimmers and Remotes
- Enerwave Switches, Dimmers and Outlets
- Evolve Switches, Dimmers and Outlets

Thermostats
- Nest Learning Thermostat
- Honeywell Lyric Thermostat
- Honeywell Total Connect Comfort Thermostats
- Ecobee3 Smarter Wifi Thermostat
- SensiWiFi Programmable Thermostat
- Haiku Home Ceiling Fans
- Keen Home Smart Vents

With that knowledge, it's time to finally learn how to actually connect your Echo with those devices!

- Open up your Alexa app and click on the three bars in the top-right corner.

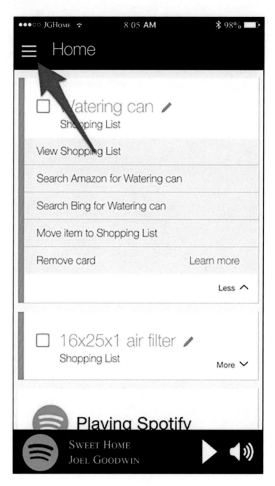

Amazon Echo in 1 Hour

- Next, a menu will appear where you can click on Smart Homes

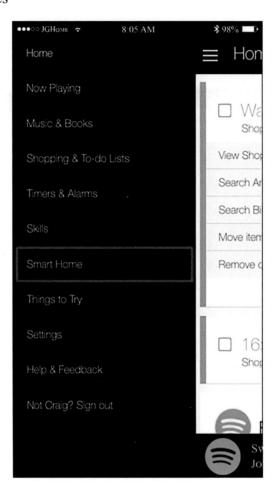

- Once there, look for the option called "Discover Devices", and click on that

Amazon Echo in 1 Hour

- Follow the menu where you will see a list of the discovered automated devices available for Echo to connect with. Just choose the device you need, and you are done!

An alternative way to add your devices

If the previous steps mentioned are giving you a hard time, it means you will need to download a new skill to get your job done!
- Go to the Smart Home screen again and click on the "Get More Smart Home Skills" button.
- Inside the Search Skills Box, type in the brand of your smart device you are trying to pair up. In this case, we are going for SmartThings and ecobee.
- Once you have found them, enable them and follow the instructions to insert the required credentials.

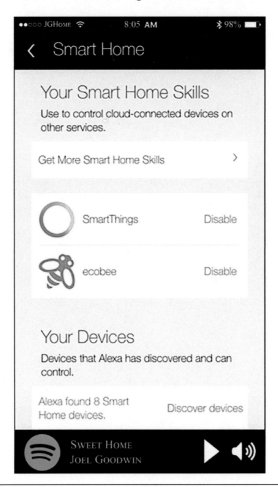

Amazon Echo in 1 Hour

- You should now be able to look for your devices.

The Final Step

- Once everything is done, you should now be able to control everything around you. But just for your added convenience, it's better to add everything in a single room to a specific group.
- For example, here we have grouped up all the lights in our bedroom into a group called "Bedroom." How will this help? You will see in a bit.

Now you should be ready to use your automated devices! Just follow the voice commands below to manipulate everything:

"Alexa, turn [groupname] [on/off]."

"Alexa, turn [on/off] [groupname]."

"Alexa, start [groupname]."

"Alexa, set [groupname] to [X%]."

Always keep in mind that you can go for the different names of your device as well to interact with them, but the group name will just allow you to make things much more accessible to you.

Chapter 5: A brief troubleshooting guide and FAQs

No electronic device in this world is perfect and everything is prone to a little malfunction every now and then. This chapter will explain some of the more commonly known issues of Echo and how you will be able to fix them.

Keep in mind that most of these issues might be solved through new updates and performance enhancements, but it's always better to stay prepared than to falter later! Let's get right into it.

Lost Wi-Fi connection

Sometimes you might face a problem where your device is unable to connect to your Wi-Fi or is being disconnected. This can be understood by the ring of light on top of your Echo exerting an unblinking orange color. Alternatively, if the light is indicating a white color, it means it is properly connected. So, if the former happens, just go through the following steps:

- Gently disconnect your device from its power cord and keep it like that for 3 seconds, and then connect it to the device again
- Then, make sure your device is properly linked to your Amazon account, and using your Alexa app, navigate to the section where you are asked to manage your devices and choose the name of your device
- From that menu, unregister your device and start up the registration process from the beginning

Unable to find connected home devices

If you are having trouble locating the devices in your home that you have connected to your Echo, you will need to go through the following steps:

- Make sure your smart device and Echo are on the network
- Update your Wi-Fi network, and then you should be able to find the smart devices

Alternatively, another method of solving the problems would be to go through the following settings:
- Utter the words "Discover my devices" while making sure your devices are set in the same band of Wi-Fi frequency as your Echo. Keep in mind that your Echo emits a frequency of 2.4 GHz as opposed to 5, so your devices should also be on 2.4 GHz
- Make sure to do that by changing your network router to 2.4 GHz and enabling SSDP or UPnp
- This should solve your problem
- Restart your Echo and the smart devices

Soft-resetting your Echo

If you are facing a scenario where you are going through a problem you are unable to solve and want to completely reset your device, go through the following steps and it will hard-reset your device:

- Note that there is a hole situated at the base of your device; take a small clip and push it through the hole and hold it there for about 5 seconds
- The light ring of your device will change its color from first orange to blue, indicating that the reset is done
- The light will turn off and then on again.
- The light will turn orange yet again, letting you know that it is now in the starting setup mode
- Here, you are going to need to fire up your Alexa app and register your device and link it to your Amazon account yet again

Echo not being able to hear properly

Sometimes you might face a scenario where your Echo will not be able to hear what you are saying. Under such circumstances, you can go for the following solutions:
- Firstly, simply go to restart your speaker
- Keep in mind that the device should be placed in a scenario where it is 8 inches away from any form of obstruction from any side
- Keep it away from anything that produces ambient noise, such as an air conditioning unit
- You can also download a skill called Voice Training, which will alter the software of Alexa, allowing it to understand 25 common phases using a typical voice the Echo will understand better

FAQ Section

This section is going to cover some of the common questions asked by new users of the Echo.

What are the most basic steps of using Echo?

If you want to ignore all the complicated steps of the advanced techniques and just want to experiment with the device, then the only thing you need to remember is, upon setting up your device, you have to say "Alexa" followed by your command to allow Alexa to perform a specific action.

Is it possible to improve the voice recognition services of Alexa?

This is a common question that sometimes floats around the minds of many people. They often ask the question, "Will the voice recognition capabilities of Alexa increase overtime?"

The answer to that question is yes. Over time, Alexa will start to process your voice recordings and train itself to understand better what you are saying. Alternatively, when you are using third party software, Echo will also extract information from those as well to train itself to understand better and process your requests based on your preferences, such as with your musical tastes.

Is there any record/history of commands given to Alexa?

It is in fact possible to go through each and every one of the voice interactions you had with your Echo by simply going through the history settings of your Alexa app. The interactions are even placed in specific groups depending on the type of your request or question to make it easier for you to find out. Just locate the command you are looking for and press the icon on the command to get more information.

Is it possible to delete the saved recordings?

Going through the history of your device through the Alexa App, you will be able to delete the specific voice recordings linked to your personal account. For this, all you need to do is go to your history and select a specific entry, tap on the delete icon and the entries you selected will be omitted.

Keep in mind that unless you are facing some security issues, you should not delete previous data, as it may drastically decrease your experience with the Echo, as Echo processes the recorded data to upgrade its voice recognition capabilities further and enhance your experience.

What are Amazon Skills?

Skills are basically apps for your Echo you can interact with using your voice. These bring about additional capabilities that allow your Echo to enhance your experience with it. The way Echo will interact depends on how the skill you have downloaded is designed. If you have a gaming skill, then Echo might indulge you in a quiz competition, or if you have Yoga, it might help you work out.

How does Echo process the shopping feature?

It is possible to order various Prime products from Amazon Prime, given that you are a Prime member. However, you are not only limited to prime products, as Alexa is designed to be able to order various normal products as well as audio books or digital musical tracks. Using the default payment and ordering mechanism, you will be able to order your products, get an audible confirmation code and check all of the information about your product through your Alexa App.

Is it possible to turn off the purchasing feature?

To prevent accidental ordering of products, it is actually possible to turn off the Voice Purchasing settings. Just go to your Voice Purchasing settings and turn off the feature from there. You will also be able to inquire about a specific order code, which will act as a security measure. No one will be able to order any product without knowing the code.

Conclusion

Now that we are done with everything, it is time to end our journey here!

Given that you have properly gone through all the instructions I had set in this book, you should have a strong grasp of both the basic and advanced functionalities of the fabulous Echo device! Be warned, though: this book is nothing but the tip of the iceberg here, and there are actually thousands more functionalities for you to explore for yourself!

With more than 1500 skills out there, and with even more coming each and every day, there is no limit to what an Echo device can do. As long as a job can be done using only audio, rest assured that Echo can do it.

From this moment on, I would like you to experiment with new possibilities whenever you can. Go ahead and turn your house into an automatic home in no time and watch as your dreams of the future come to reality.

I would like to thank you for purchasing this book and I do hope it was an informative and fun read for you.

Good day and God bless!

Made in the USA
Middletown, DE
26 December 2017